YOUR KNOWLEDGE HAS VALUE

- We will publish your bachelor's and
  master's thesis, essays and papers

- Your own eBook and book -
  sold worldwide in all relevant shops

- Earn money with each sale

Upload your text at www.GRIN.com
and publish for free

Heinz Rüterbusch

High level technical design and economic assessment of renewable energy solutions for radio base stations

Nutzung von erneuerbaren Energien für Mobilfunkstationen

GRIN Verlag

Bibliografische Information der Deutschen Nationalbibliothek:

Die Deutsche Bibliothek verzeichnet diese Publikation in der Deutschen National-
bibliografie; detaillierte bibliografische Daten sind im Internet über http://dnb.d-
nb.de/ abrufbar.

Imprint:

Copyright © 2008 GRIN Verlag GmbH
Druck und Bindung: Books on Demand GmbH, Norderstedt Germany
ISBN: 978-3-640-19621-0

This book at GRIN:

http://www.grin.com/en/e-book/115425/high-level-technical-design-and-economic-
assessment-of-renewable-energy

High level technical design and economic assessment of renewable energy solutions for radio base stations

by

H. Rüterbusch

Dipl.-Ing (FH) / Dipl. Wirt. Ing. / Dipl.-Kfm

Term paper on the subject of 'autarkic energy systems' as part of the University of Kassel/Germany's renewable energy study module

Hausarbeit zum Schwerpunkt "Autarke Energiesysteme"
im Rahmen des Studienmoduls "AEE" des Studiums "E+U"

TABLE OF CONTENTS

U N I K A S S E L
V E R S I T Ä T

1 Introduction

1.1 About this document

This document describes a high level technical design for and the economic assessment of renewable energy solutions for radio base stations (RBS). It proposes a way forward in the development of 'hybrid' solutions with renewable energy sources (RES) for off-grid RBS sites.

The main purpose of this document is to provide initial practical, technical and economic guidance. It is accompanied by an easy-to-use EXCEL-based tool (RAFORS – Renewable energy Assessment FOr Radio Sites) for the assessment of the net present costs (NPC) of these solutions.

The following sections outline the specific requirements of mobile operators and provide specific technical guidance for the selection and implementation of the elements of a hybrid system.

This information will enable mobile operators without any detailed technical and commercial knowledge to develop initial ideas on the design of such a system and provide them with a common basis for further discussion, evaluation and benchmarking.

References to external documents in the glossary are in square brackets [ ] and cross links to specific sections of this document are indicated as follows: >> section_name.

Note: this document is part of my final examination in the renewable energy study module at the University of Kassel. The precise definition of the term paper can be found in >> Issue definition

1.2 About the author

Comments, remarks and questions about this document are welcome! Just send an e-mail to

Heinz.rueterbusch@gmx.de

The author is working in Technology Network Department of an international mobile operator on energy and Infrastructure solutions to support the operating companies.

He is a member of the Environmental Engineering (EE) Group within the European Telecommunications Standards Institute (ETSI).

1.3 Issue definition

Subject: <High level technical design and economic assessment of renewable energy solutions for radio base stations>

The aim of this paper is to describe and discuss the different aspects of the technical design and economic assessment of renewable energy solutions for radio base stations. The following points should be considered:

A.   Load profiles, user requirements, resources, typical technical concepts for radio base stations. >> Technical characteristics

B.   Renewable energy-oriented technical design. Possible concepts, adaptation of different energy sources, integration of the components (PV, wind generator, diesel, batteries, chargers/inverters, management/control, accessories) >>How to get RES@ and >> Technical design

C.   Economic assessment and comparison of different technical solutions >> High level RES assessment (RAFORS)

D.   Design of a "Site questionnaire to plan RES based radio stations" >> Example of a 'Site questionnaire to plan RES based radio stations'

E.   Presentation of an example (including schematic diagram and nominal characteristics of the components) >> Example of an RES@RBS assessment with RAFORS

Note: >> indicates the section in which the aspect is covered.

1.4 Scope

1.4.1 In scope
- Off-grid RBS sites
- Renewable energy
- Technical aspects
- Economic assessment

1.4.2 Out of scope
- Detailed site-specific planning
- Switching centres
- Specific suppliers

2   Mobile operators

2.1   Role and motivation

Two of the key challenges facing mankind today are protecting the environment and combating climate change. Mobile operators are part of the information and communication technologies (ICT) sector and, as such, they play a key role since they are both part of the solution and part of the cause.

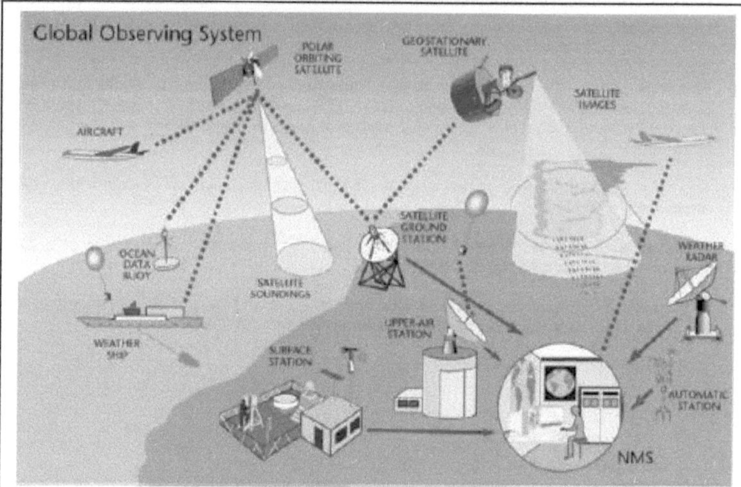

Figure 4: WMO Global Climate Observing System
Source: World Meteorological Organisation
Note: NMS = National Meteorological Service.

Figure 1: Example of how ICTs are contributing to global warming [1]

The illustration shows how telecommunication services can help to prevent global warming by providing an observing system.
ICTs consume power, radiate heat and emit carbon dioxide due to the operation of the telecom network >> Technical context. The objective is to reduce carbon dioxide emissions because these emissions are what cause global warming. A reduction of carbon dioxide emissions can be achieved principally by

- improvements in energy efficiency and
- increased use of renewable energy sources (RES).

The challenge is to achieve energy-efficient and sustainable mobile communications through network and site optimisation and the use of renewable energy sources.

The use of RES will also help to reduce future operating expenditure (OPEX) on energy. However, the full net present costs (NPC) and financial expectations also have to be taken into account when considering alternatives.

2.2   Technical context

The technical devices that are necessary for the operation of a mobile-to-mobile connection can be logically assigned to a 'core' network and an 'access' network. The access network contains the radio base stations (RBS), which are also called sites or base transceiver stations (BTS).

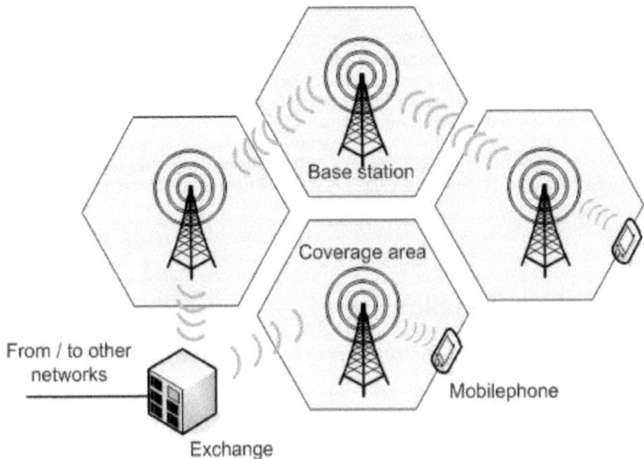

Figure 2: How a mobile network works, similar to [8]

The exchange switch is part of the core network. Around 80% of expended energy and the related carbon dioxide emissions are accounted for by the access network due to the number of RBS sites. Today's European networks operate two technologies concurrently: 2G (GSM) and 3G (UMTS).

The RBS equipment mainly operates on direct current power (DC), with the exception of the conventional air-conditioning (AirCo or cooling) systems which cool the equipment and batteries >> Batteries. The following table shows the items of RBS equipment and their typical power consumption.

RBS equipment	Consumption	%	Type	Remark
Radio equipment	2000W	62%	DC	2G+3G (co-located)
Transmission equipment	300 W	9 %	DC	Links and nodes
Active cooling	750W	23%	AC	Measurements
Free cooling	75W	2%	DC	
Other	125W	4%	AC/DC	O&M, -48V PSU, …
Total	**3250**	**100%**		Including cooling

Table 1: Power consumption at a co-located radio site

RBS sites mainly require DC power and the radio equipment has the highest power consumption.

Due to the fact that that RBS operate 24h/365d and the impact on traffic variation is less than 20%, an almost constant load throughout the year is assumed.

Unlike radio equipment, cooling systems are not operated 24h/365d. However, as a result of outside temperature fluctuation, it is practical to evaluate load requirements in the summer and winter seasons and to take the worst case (highest value). Some hybrid planning software such as HOMER (>> Site perspective: planning steps) can define daily consumption profiles for different DC sources.

Since it is assumed that mobile operators will be able to reduce the use of conventional cooling systems, the RAFORS model only takes DC load into account. This is possible because the ETSI EN 300 019 [1] environmental standard allows the environment in which the equipment is installed to vary. For example, in class 3.1 room temperature can vary from 5°C to 40°C and in class 3.2 (partly temperature controlled) room temperature can vary from -5 to +45°C [3].

Technological advancements which boost the efficiency of radio equipment will lead to lower power consumption in future, and the typical power consumption of an RBS is expected to decline to around 2 - 3kW.

2.3   Technical characteristics

The following section outlines the key technical characteristics of RES applications which differ from island applications such as small villages:

- Mainly DC loads, operating 24h/365d

- Most common DC voltage is -48V

- Relatively small loads (less than 5KW on average, typically 2 - 3 KW)

- Relatively stable and predictable long-term loads (through the use of new technologies)

- High numbers of radio sites (e.g. Vodafone has around 85,000 sites [8] worldwide)

- Space restrictions, especially in urban areas

Several technical solutions (e.g. grid, diesel engine etc.) are available to power radio sites, so it is important that an RES solution has a comparable total cost of ownership (TCO), expressed in terms of net present costs (NPC).

These circumstances make the application or evaluation of renewable energy for pure off-grid sites more feasible since off-grid sites

- have less space restrictions (e.g. for photovoltaic (PV) modules)

- enable clear benchmarking with costs of diesel engines

- have long maintenance intervals (no fuel logistics)

Aspects such as the solar panel theft and the incidence of the renewable energy sources have to be considered on a case by case basis.

Maintenance and site visits also have to be made to off-grid sites at least once a year.

3   How to get RES@RBS

3.1   Network perspective: prerequisites (macro view)

Energy optimisation in mobile communications is a step-by-step process, starting with an energy efficient network design followed by an energy efficient RBS site design [3].
An energy efficient site and network design (macro view) - also in terms of legacy elements - can be achieved in the long term in the following order by

1. Designing to real needs

2. Optimising the existing infrastructure or radio equipment

3. Introducing more efficient technology

4. Considering RES as a potential power source (site specific, micro view)

Guidance on the first three bullet points already exists in the form of the ETSI's Technical Report on 'The reduction of energy consumption in telecommunications equipment and related infrastructure.' [3}. This document is an accumulation of ideas from operators and manufacturers on the methods to increase the energy efficiency of telecommunication systems in order to reduce operational energy requirements.

For example, the introduction of higher temperature set points enables the replacement of conventional 3-phase cooling units with free cooling ventilation units which consume less power and operate on DC power. Please remember, however, that elements of the design to real needs phase may have an impact on the entire network. It could take quite some time to complete since existing operations with related organisational processes are affected.

When site optimisation has taken place, RES can be considered from a site-specific perspective.

3.2   Site perspective: planning steps (micro view)

The following illustration shows one possible approach from the assessment phase to the site-specific implementation of a renewable energy solution. The level of detail increases progressively over time from left to right.

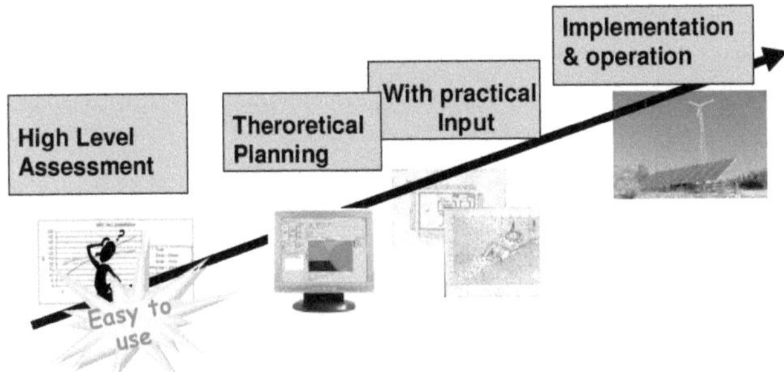

Figure 3: The 'RES@RBS' planning concept

Off-grid systems, particularly hybrid systems, are characterised by a high degree of complexity at the dimensioning stage. That's why software simulation tools are extremely useful. [13]. The term 'hybrid' system in this document refers to the elements covered in >> Selected technical recommendations

One possible initial assessment tool is discussed in >> High level RES assessment (RAFORS). Several tools for the other introduction phases are also available. Planning software for full hybrid solutions can be placed in the categories of

- dimensioning programs (Dim), which calculate the system dimensions on the basis of input data (load and climate data and system components), and

- simulation programs (Sim), which use the input data (load and climate data, system components and configuration) to simulate the behaviour of the system over a given period.

A good summary is provided by [13].

The following table shows the software tools used in the planning phase.

RES	Software	Source	Remark
Solar	Sunny Design	http://www2.sma.de/en/solar-technology/downloads/index.html	Solar module database included
Wind	Wind Pro	http://www.emd.dk/	
Hybrid	HOMER	https://analysis.nrel.gov/homer/default.asp	Dim/Sim, calculates operating costs
Hybrid	RETScreen	www.retscreen.net/ang/t.php	Based on EXCEL, Dim

Table 2: Example of planning software tools

One of the key decision criteria for users selecting hybrid tools is the kind of calculations that the tools can make: economic calculations (HOMER), general dimensioning (RETScreen) or a detailed technical configuration. Alternatively, the planning can also be performed manually in the following logical order:

1. Assessment of resources (wind, solar)
2. Dimensioning of the PV generator, wind generator (power and type)
3. Battery dimensioning
4. Selection of charger and inverter

Many of the software tools include wind or solar databases, so any site questionnaire should include the geographical location of the site >> Example of a 'Site questionnaire to plan RES based radio stations'

The first consideration should be solar energy (photovoltaic, PV), because

- PV energy yield is easy to predict (solar radiation)

- PV systems are simple to install and maintain and therefore

- PV systems reduce operational costs to the minimum.

The following sections provide an overview of the components in a hybrid system and the prediction of natural resources.

4   Selected technical recommendations

4.1   General

The following section provides guidance on the selection and design of the technical elements. This document does not recommend a specific vendor. However, for practical reasons a number of suppliers and products are presented in the annex >> Suppliers as an example.

> Initial pointers on the manual assessment of solar and wind resources are provided at the end of the wind and solar section.

4.2   Technical design

Although the load requirements are mainly DC, the air-conditioning system may require some AC. Diesel generators are also used as a back up or main power source at most sites. As a result, a 'mixed' design has to be applied. The diagram below shows the two possible designs.

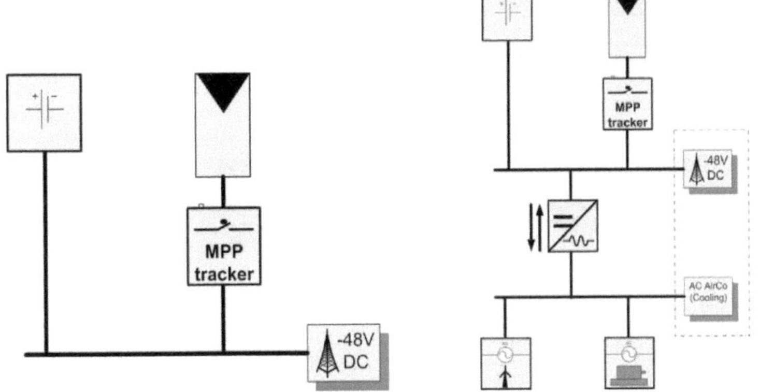

Figure 4: Recommended technical designs: DC only and mixed (AC and DC)

The power generators are on the left hand side of the diagrams and the BTS load is on the right hand side in the shadowed boxes. A key to the symbols is provided in the annex > Symbols.

The bi-directional battery charger is a bottleneck in the 'mixed' design on the right hand side. However, this might be an acceptable drawback since relatively stable or even decreasing load demand is expected in future >> Technical characteristics.

Both designs can be adapted. For example, it would be possible to start off with the mixed design, which allows diesel generator (DG) backup, and phase out the DG at a later time. This evolutionary process is shown in the following diagram:

Possible evolution of a green site

Flexible design of a Hybrid follows operational experience

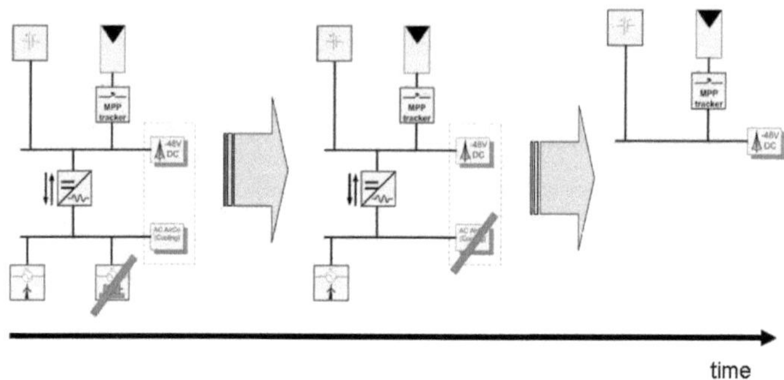

Figure 5: Evolution of a green site and impact on the technical design

If it turns out not to be necessary to operate a DG in parallel with the RES after a while, the DG can be taken out of the scenario and redeployed elsewhere. It may be that there are no requirements of active cooling after some time, or the need for a wind turbine is reconsidered. The end result would be a purely DC based design.

The design is therefore flexible with regard to

- DC power requirements

- AC power requirements

- backup source (e.g. diesel or fuel cell)

The following sections describe the individual elements of the design.

4.3 Solar

The following considerations are important when selecting a PV module [7. 8}:

- Price per Wp (http://www.solarbuzz.com/)

- Performance tolerance (3% - 5% recommended)

- Warranty (typically 20 years, depending on TCO)

- Efficiency (approx. 15% to 18% for monocrystalline units)

When comparing technical attributes it is important to establish whether a cell or a PV module is being referred to. A module consists of cells and often comes with a physical frame.

PV generators produce energy from diffuse and non-diffuse solar radiation. In many cases, therefore, no solar tracking system is necessary (tracking systems consume power and require maintenance). The following table shows the space requirements of the most common PV cell technologies:

Type of PV module	Space required for 1kWp	Remark
Monocrystalline	7 – 9 sqm	8 sqm default in RAFORS
Polycrystalline	9 – 11 sqm	
Amorphous	16 – 20 sqm	

Figure 6: PV module floor area requirements

Amorphous PV modules are the least expensive option. Shadowing, even only on part of the module, will dramatically reduce energy output [7, p.138}. If shadowing is unavoidable, shadow-tolerant modules or the installation of a higher number of modules may help.

The higher the temperature of the PV module the lower the energy yield. The Standard Test Conditions specify 25°C as the maximum limit. Each degree of higher temperature reduces PV output by 0.5% [6, p. 17]. Effective rear ventilation of the modules is therefore recommended.

Solar radiation measurements can be found on the local meteorological services' websites. Meteonorm (Switzerland) also has a European Solar Radiation Atlas, E.S.R.A. - a very broad database: www.meteotest.ch/

4.4 Wind

Wind turbine generators (WTG) are more efficient (about 40%) than solar power, but a mast is needed to operate them, which leads to higher operating costs. The main parameters for wind power are

- Average wind speed / electricity output

- Investment costs for the WTG and ancillary costs for foundations etc.

- Operation and maintenance costs

- Turbine lifetime

The power curve is therefore an important technical criterion for WTG performance. An example is shown below.

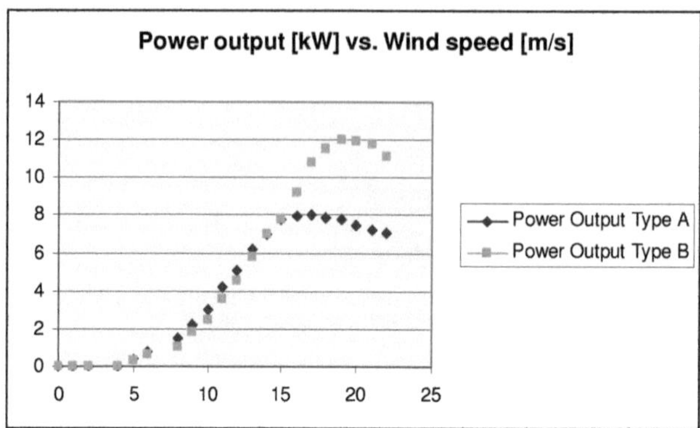

Figure 7: Example of a power curve of a wind turbine generator

Conventional turbines are efficient at wind speeds of between appr. 10m/s and 25m/s [10]. In combination with the wind measurements over time, the curve is an important technical input parameter for the prediction of the WTG's energy consumption.

It is also necessary to consider the higher maintenance requirements of WTGs which affect OPEX. The OPEX-related factors to be considered in connection with WTGs are listed below.

Factor	%
Land rent	18
Insurance	13
Administration	21
Power from grid	5
Service spares	26
Miscellaneous	17

Table 3: Percentage of OPEX spent on wind turbines [10]

Some WTGs require external power to operate. OPEX comprises around 10% of the investment costs and shown in >> Wind. OPEX increases proportionately to length of operation. After 10 years of operation, larger repairs and extensive maintenance may be necessary and, based on experience, these are the dominant costs during the last 10 years of turbine life [10].

A WTG may need an additional inverter for connection to the AC or DC bus since the wind turbines produce unregulated AC.

Wind measurements should be made over a minimum one year period. An initial overview of the wind situation can be found at several websites such ashttp://www.winddata.com/.

4.5 Batteries

Lead-acid batteries are the most common type of battery. They are used in automobiles and electricity consumers as a backup energy source. A couple of variations on the traditional design have emerged:

- *Valve-regulated lead-acid (VRLA) batteries* — are sealed and need no topping off with water. This type requires less maintenance than regular lead-acid batteries.

- *Gel-type lead-acid batteries* — are filled with a gel instead of liquid. This makes them much less likely to spill.

The lifetime of a battery depends on the number of capacity and the depth of discharge (DoD). OPzS or OPzV types of accumulators are designed for higher number of charge and discharge cycles. Thus these batteries are recommended for 365 days a year radio base station applications. The OPzV variant with fixed electrolyte (gel) should be applied if site ventilation is not optimal and to reduce maintenance requirements [4].

The following picture shows the DoD as a % of the number of cycles of OPzV batteries:

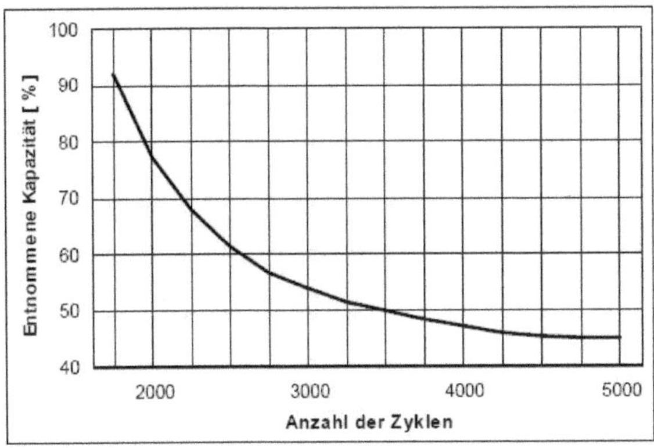

Figure 8: Extractable energy versus number of cycles of OPzV batteries [4]

The RAFORS tool assumes 40% DoD and the use of OPzV/S batteries to maximise the lifetime.

A high temperature reduces the lifetime of lead acid batteries. Battery output can be sustained at RBS sites if the batteries are cooled separately in case that active cooling is not applied > Technical context. Alternative solutions are provided by [3].

4.6   Charger, controller

There are two kinds of controller:

- solar charge controller to charge batteries from PV power

- stand-alone inverter as a bi-directional battery charger if AC is used

The solar charge controllers are available with or without maximum power point (MPP) tracking for the PV generator. The efficiency of the solar charge controller without MPP tracking is over 99% and will be less with MPP tracking. An inbuilt temperature sensor should be used to measure the temperature close to the batteries to ensure efficient charging >> Batteries.

The solar charge controller should therefore be located close to the batteries [4]. The following illustration shows an example of a circuit diagram in a DC only design >> Technical design

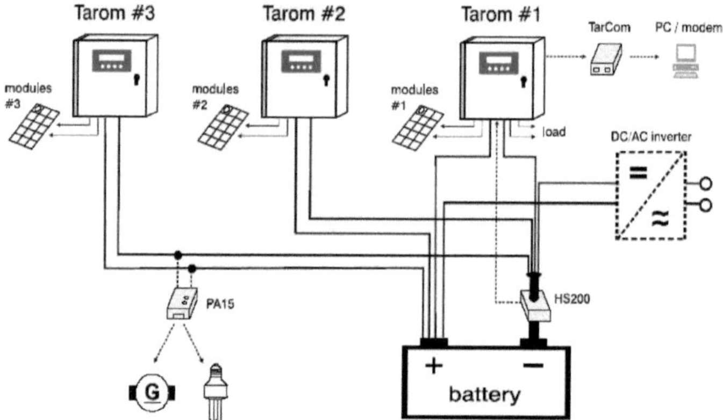

Figure 9: Example of the parallel operation of a charge controller [9]

The load can also be connected directly to the DC bus as shown in >> Technical design. In this case a different charging algorithm based on voltage will apply. The DC/AC inverter is not necessary for pure DC loads.

If an AC voltage grid is needed (e.g. if the site has electricity generators or AC-based air-conditioning systems) a bi-directional battery charger as a stand-alone inverter is recommended. The efficiency is about 95%. Central logic is necessary to control the diesel GenSet start-up> Diesel generator. This could be provided by the existing inverter / controller so that an additional controller is not necessary at the site.

The charge controller should also be capable of parallel operation in case higher loads are needed. An example of a 3-phase application with DG control is shown in the Annex >> Example of a 3-phase design.

4.7   Diesel generator

A diesel generator (DG or GenSet) is often used as a back-up system in a hybrid design. Due to the high OPEX and CO2 emissions, it is recommended that the generator is only used at selected sites.

For example, it can be used at the outset and then, if measurements prove that it is not necessary, it can be phased out. It may be necessary to optimise the hybrid design until generator running time is zero. De-installation is possible due to the flexible design > Technical design

The DG 'stands in' as a

- back-up power source or
- to cover peak demands for AC power.

DGs are typically most efficient at between 80% and 100% of full load. Power of between 10KVA and 20KVA are applied in the mobile communications sector. The effective GenSet power in KW can be calculated by multiplying the idle power, measured in [kVA] with cos phi (about 0.8). Example: 25 KVA is around 20 KW.

Diesel GenSets rarely operate at the full load for which they are designed [4].

So it is important to ensure that the GenSet is only used if the batteries cannot cope with the load demand, so intelligent control is necessary. It should be possible to control the DG, e.g. via external input signals to 'start' and stop' the engine.

The photo below shows two DGs.

Figure 10: Diesel GenSets (Source: SMA [15])

The DGs are shown without the fuel tank.

The nominal power of the GenSet should be adapted to the nominal power of the bi-directional charger.

4.7.1 Accessories (cabling)

One accessory is DC power cabling. A rough calculation of required cable diameter can be made on the basis length and applied power: [4]

$d = 0.0175 \times L \times P / (f_k \times 48V)$

Where

A = cable diameter in mm^2

0.0175 = the specific resistance of copper [Ohm x mm2 / m]

L = cable length, taking positive and negative wire into account in m

P = power in W

f_k = insertion loss factor, 3 % is recommended = 0.03

RAFORS considers all losses, like cable insertion losses, in a quality factor 'V' >>Table 6: Default hybrid scenarios and corresponding parameters in RAFORS

5 High level RES assessment (RAFORS)

5.1 Purpose

The EXCEL-based software tool is easy to use and should enable telecom operators to perform the first technical and economic assessment for renewable energy solutions (RES).

The intention was to design a transparent, understandable and easy-to-use tool to provide

- initial technical and economical figures

- a common basis for further discussion and evaluation

- a tool to crosscheck / compare calculations.

RAFORS enables benchmarking by comparing with an e.g. outsourcing of a 24/365d running DG to power sites.

5.2 Abstract

RAFORS is a user-friendly tool for the initial technical assessment and economic analysis of a defined range of hybrid power systems for radio base stations. Thus it is only necessary to enter a couple of input parameters so that the tool can provide initial technical and economic output for further evaluation and discussion.

It is based on EXCEL to ensure ease of use, adaptation and simple comprehension of the spreadsheets. The >> Abbreviations (formulas) section also provides an overview of the abbreviations used. RAFORS offers two basic methods of assessment:

- a quick check of 4 predefined hybrid combinations with only 3 data input entries or
- a self-definable hybrid design based on more selectable data input on loads, wind speed, etc.

Please note that the current version 1.0 only has the Quick Check function for the predefined combinations. The following illustration shows the predefined options represented by symbols.

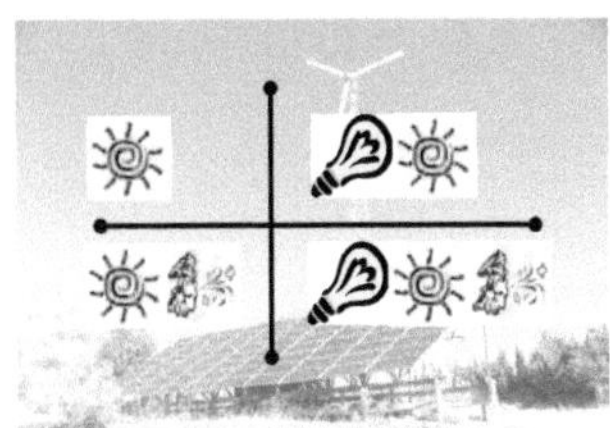

Figure 11: Predefined hybrid combinations in RAFORS

These symbols also identify the respective input / calculation area in the Calculation spreadsheet, >>Figure 14: Screenshot extract of the calculation spreadsheet in RAFORS

The tool was designed to assess a limited range of hybrid power systems. More complex planning tools such as HOMER should be used for later planning purposes. These tools are used to define the

final technical design when the resources have been assessed >>Site perspective: planning steps (micro view).

A first economic analysis for the operation of the above-mentioned solutions over a six-year operational period is included (implementation beginning of year 0 and 5 years operation). It calculates the economic value of the project on the basis of interest costs and average financial input costs.

5.3 Simplifications / restrictions

The simplifications in the following table have been made to ensure that the model is comprehensible:

Simplification	Reason
DC only	Radio sites mainly use DC power, free cooling is also DC-based
Constant load per day	Radio sites are operated 24h/365d like Pmax
Annual average solar radiation, wind resource	For an initial assessment
Physical constraints (site size etc) are not taken into consideration	Next planning phase
Selected fixed system configurations	No planning!
ROI in 5 years operation	fast payback, no re-deployments necessary

Table 4: Simplifications in RAFORS

The calculation of AC (wind, DG) requirements is more static because wind and AC load requirements (e.g. air inflow in air conditioning systems) are much harder to predict.

The following symbols are used to identify components (or the scenario):

Symbol	Hybrid element
	Solar, PV
	Wind, WTG
	Diesel GenSet (/load)

Figure 12: Power source-related symbols in RAFORS

The light bulb at the top right hand corner of the spreadsheet represents load and photos are used to help identify elements/scenario.

5.4 How to work with RAFORS

5.4.1 General

RAFORS has three EXCEL spreadsheets

Name of spreadsheet	Content	Remark
Output	Overview of output	Input for DC load and benchmark
Calculation	Reaming input data	'Heart' of the tool
Wind	Supporting assessment of wind incidence	energy production based on rotor diameter

Table 5: RAFORS spreadsheets

The wind spreadsheet is not a focus of the following sections to keep the document short. Each spreadsheet has the following cell colour coding:

green	Input technical commercial
red	calculated oι fields
grey	pre-filled remarks and data
dark green	do not change

Figure 13: RAFORS colour codes

Data is only entered in the light green field. The sheet itself is protected to allow version-controlled release management.

An example of an assessment can be found in >>Example of an RES@RBS assessment with RAFORS. Formula abbreviations are shown in >> Abbreviations (formulas)

5.4.2 Input data

The DC load of the BTS and the 'benchmark' costs can be entered in the output spreadsheet, while all other input variables and calculations are done in the calculation spreadsheet:

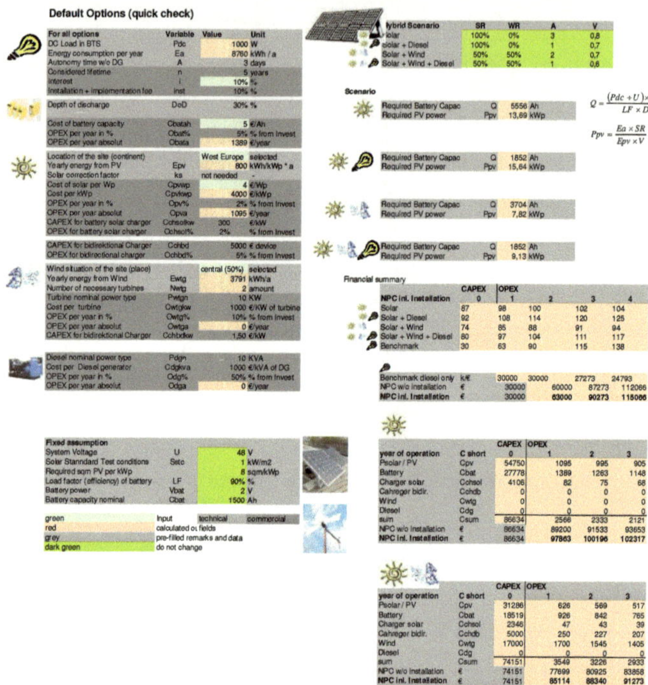

$$Q = \frac{(Pdc + U) \times}{LF \times Dc}$$

$$Ppv = \frac{Ea \times SR}{Epv \times V} \, ;$$

Figure 14: Screenshot extract of the calculation spreadsheet in RAFORS

The input values can be edited on the left hand side and the calculations of the four options are made on the right hand side. The formulas are explained in >> Calculation

5.4.3 Output

The output summary is displayed in the overview spreadsheet. To convert the number of batteries for each scenario please refer to >> 5.5.3

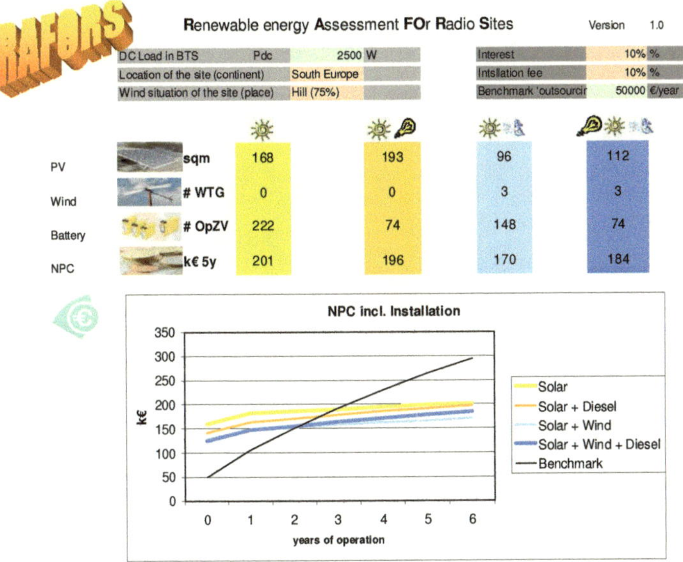

Figure 15: RAFORS output summary screenshot

Each element's output and NPC for 6 years' operation are displayed. It is assumed that the CAPEX is spent at the beginning of year '0' and that operational costs are incurred after one year of operation (at the end of that year).

5.5 Calculation

5.5.1 Power

Energy consumption per year in [kWh/a] is based on the average DC power of an RBS

$$Ea = Pdc \times 365 \times 24 \div 1000$$

5.5.2 PV

The formula follows assessment approaches such as [15, 4], considering a Solar Ratio SR depending of the hybrid scenario >> Table 6: Default hybrid scenarios and corresponding parameters in RAFORS

Required PV power [KWp]
$$Ppv = \frac{Ea \times SR}{Epv \times V} \approx \frac{Ea \times SR \times Sstc}{365 \times Sxo \times ks \times V}$$

RAFORS version 1 uses the energy gain Epv per year formula on the left. The more detailed formula on the right will be used for specific design in RAFORS version 2. The values of V and Epv are taken from [15].

The solar ratio SR and system efficiency V depend on the hybrid model and represent the complexity of the system. Please refer to the 'calculation' spreadsheet.

Hybrid Scenario	SR	WR	A	V
Solar	100%	0%	3	0,8
Solar + Diesel	100%	0%	1	0,7
Solar + Wind	50%	50%	2	0,7
Solar + Wind + Diesel	50%	50%	1	0,6

Table 6: Default hybrid scenarios and corresponding parameters in RAFORS

The table shows that the introduction of a diesel generator reduces the necessary capacity of the batteries to 1 day autonomy. Note: the ',' stands for '.' In the column 'V' (not written in British English)

The solar factor (ks) reflects the conversion of the solar energy into energy which can be gained from PV (Epv in kWh). A default figure of 0.8 can be assumed. The energy delivered by a solar generator also depends on factors relating to its physical location such as

- temperature impact (ktemp)
- horizontal and vertical physical alignment of the PV module (kalign)
- the nominal peak power of the PV module

$$Epv = Pnom \times \frac{Sxo}{Sstc} \times ktemp \times kalign$$

The formula uses the nominal power of a solar module in kWp. Please check whether Sxo refers to a daily or annual value!

A module surface of 8 sqm per kWp has been assumed in the output sheet >> 4.3 Solar

5.5.3 Battery

The formula follows assessment approaches such as [15] for a C10 battery type.

Required battery capacity in [Ah] $Q = \dfrac{(Pdc \div U) \times A \times 24}{LF \times DoD}$

The DoD depends to a great extent on battery type and charging cycles >>Batteries. The model assumes 40% taking into account 1 cycle of 10h per day (night / day).

The number # of batteries in the output sheet is calculated by

$$\#batteries = \dfrac{Q \times U}{Cbat \times Ubat}$$

By default, RAFORS assumes 2V OPzV/S batteries with a Cbat of 1500Ah. The system works on a voltage U of 48V.

5.5.4 Wind

The wind energy calculation in the model is more or less static since wind energy is not easy to predict. An approach which assesses the energy based on rotor diameter has been chosen to keep the model and the required input data simple. Please refer to sheet 'Wind' in RAFORS.

The input is provided from WTG specifications and real energy yields. Then an 'iterative' assessment is made to provide an idea of annual wind energy production levels.

The calculation uses a formula to assess wind power in kW based on wind speed and rotor diameter:

$$Pwtg = 0{,}5 \times Da \times v^3 \times \dfrac{d^2}{4} \times \pi \times fw \div 1000$$

$$Ewtg = Pwtg \times 8760 \times WR$$

The Wind Ratio depends on the scenario and accounts to 50% for wind applications in the default scenarios >> Table 6: Default hybrid scenarios and corresponding parameters in RAFORS.

The autonomy time is reduced by one day [15], since two independent and non-correlated resources (solar and wind) are available.

5.5.5 Diesel generator

The diesel generator (DG) is a static device from RAFORS' perspective. In this model, it is used only as a backup to reduce autonomy time. The scenarios with diesel GenSet therefore have an autonomy time of 1 day, assuming that the site is initially powered by batteries.

The fixed amount of 15kVA has been chosen for this scenario. 15KVA represents /1.44 kW: 12 kW. This should be sufficient to support the typical load of a BTS >>Technical context, to operate the AirCo and to charge the batteries.

5.5.6 Economic aspects

All formula and value abbreviations and symbols are summarized in the Annex >> Abbreviations (formulas)

The capital expenditure (CAPEX) and operational expenditure (OPEX) per device are indexed by 'xyz' to indicate PV, turbine, battery, charger/inverter. The calculation is made as follows:

cost = cost per unit * number of unit.

This should allow for the individual adaptation of values from technical and commercial perspectives. The unit type is also represented by a block of characteristics 'xyz' behind the 'C' for CAPEX and 'O' for OPEX figures. The following table shows the calculations.

Elements	CAPEX		OPEX (1 year)	
	C xyz	Formula	O xyz	Formula
Solar / PV	Cpv	Ppv x Cpvkwp	Opva	Cpv x Opv%
Battery	Cbat	Q x Cbatah	Obata	Cbat x Obat%
Charger solar	Cchsol	Cchsolkw x Ppv	Ochsola	Ochsol%*Cchsol
Charger bidir.	Cchbd	*Fixed assumption*	Ochbda	Ochbd%*Cchbd
Wind	Cwtg	Nwtg x Pwtgn x Cwtgkw	Owtga	Cwtg x Owtg%
Diesel	Cdg	Pdgn * Cdgnkva	Odga	Odg%*Cdg

Table 7: CAPEX and OPEX calculation formulas in RAFORS

The OPEX for the first year is displayed in RAFORS in the input area on the left hand side of the 'calculation' spreadsheet for crosschecking purposes. This doesn't apply to the OPEX in respect of chargers / inverters since this expenditure is negligible compared with the other types.

The charger types can be load charger (abbreviated to 'lc' or, if it has an AC bus, a bi-directional battery charger or stand alone inverter (abbreviated to 'bd'). The OPEX is the total investments for chargers.

The default assumptions for CAPEX and OPEX are summarised below.

Elements	CAPEX (O xyz %)		
	Recommended	Source	Used in RAFORS
Solar / PV	€ 3 – 5/Wp	solarbuzz	€ 4/Wp
Battery	€ 2 – 4/Ah	[11]	€ 5/ Ah (OpZV)
Charger solar	€ 0.8 – 1.3/W	[11]	€ 0.3/W
Charger bidirect.	€ 1 – 1.5/W	[11]	€ 1/W
Wind	€ 2000/KW	[14]	€ 1000/KW
Diesel	€ 200 – 500/KVA	[11]	€ 1000/KVA

Table 8: CAPEX default assumptions in RAFORS

Please note that the costs of installing a DG are higher since it requires a tank, canopy, etc. An additional 50 – 100 % on top of pure CAPEX [11, 12] have to be assumed.

Elements	OPEX (O xyz %)			
	Proposed	Source	In RAFORS	Reasonable
Solar / PV	2%	[11]	2%	Insurance, theft
Battery	5%	[11]	5%	Waste management
Charger solar	2%	[11]	2%	Maintenance
Charger bid.	5%	[11]	5%	Maintenance
Wind	5- 10%	[14], [10}	10%	Land rent, insurance, maintenance
Diesel	20% (w/o fuel)	[11], [5]	50%	Maint. + fuel

Table 9: OPEX default assumptions in RAFORS

OPEX, especially for wind turbines and chargers, is not distributed evenly over time and may be lower in the first years and higher in later years >>4.4. OPEX for the DG is hard to predict and should be updated based on real experience.

The final Net Present Cost (NPC) calculation involves the addition of depreciated annual OPEX to the CAPEX of year '0'. Thus NPC consists of the CAPEX of the different hybrid elements and the sum of the depreciated OPEX during the lifetime (n=6: the first year + 5 years operation)

$$NPC = \sum CAPEXxyz + \sum_{i=0}^{n-1}\left(\left(\sum OPEXxyz\right)/(1+i)^n\right)$$

Please note that CAPEX is assumed as t=0 (first year as reference) while operational expenditure is shown for 5 years .The first OPEX, reflecting one year of operation, is not depreciated because it is accounted to the reference year of the investment (year 0)

An installation fee is also added to the CAPEX (default 10%). The interest can also be included in the 'calculation' sheet.

5.6   Way forward

The RAFORS tool and the content of this document are subject to technical modifications. The intention was to facilitate discussion in a sector of industry that is not necessarily energy-oriented.

The ETSI is planning a technical report on 'THE USE OF ALTERNATIVE ENERGY SOURCES IN TELECOMMUNICATION INSTALLATIONS.'(TR 102 532)

It is quite possible that DG technology will be replaced by fuel cell systems in future [16}. The use of electrolysers and RES could then lead to energy-autonomous sites.

From a commercial viewpoint, the hidden costs of fossil fuels [7] should also be considered, however, increasing energy prices may change opinions on the economic viability of RES.

6   Definitions

6.1   Glossary

Reference	Document
[1]	ICTs and Climate Change, Watch Briefing Report No. 3, ITU-T, November 2007
[2]	Sustainable energy use in mobile communications, White Paper, ERICSSON, August 2007
[3]	The reduction of energy consumption in telecommunications equipment and related infrastructure; Technical Report, TR 102 530, ETSI, April 2008
[4]	PV-Inselanlagen - Komponenten und Auslegung, TU Berlin, Dipl.-Ing. Rainer Morsch, June 2001,
[5]	Sustainable solutions for power generation schemes for telecommunication applications in Greece, CONERGY, 2006
[6]	Planung von Photovoltaik Anlagen, F. Konrad, published by Hanser Vieweg, April 2007
[7]	Regenerative Energiesysteme, V. Quaschning, published by Hanser, 2007
[8]	The potential of communications, CR report, Vodafone, March 2007
[9]	Installation and Operation Instruction Manual, Steca
[10]	Economics of wind power, Riso institute, Denmark, 2004, ewea_050531_Economics.pdf
[11]	Autarke Energiesysteme, Script, Uni Kassel, Anlagenplaner Erneuerbare Energien, Stratis Tapanlis, >2002
[12]	Wirtschaftlichkeitsbetrachtungen, Script Uni Kassel, Anlagenplaner Erneuerbarer Energien, Stratis Tapanlis, >2002
[13]	World-wide overview about design and simulation tools for hybrid PV systems, Georg Bopp, Anja Lippkau, CONERGY, Fraunhofer Institute
[14]	Windenergie, Script, Uni Kassel, Anlagenplaner Erneuerbarer Energien, S. Heier, >2008
[15]	Grobauslegung von Sunny Island Inselsystemen, Presentation, SMA
[16}	PV-Wasserstoffsysteme zur autonomen Versorgung von Telekommunikationseinrichtungen, Fraunhofer/CONERGY, 2002

6.2   Abbreviations

AC	Alternating Current
BTS	Base Transceiver Station
CAPEX	Capital Expenditure
Cos Phi	Factor to calculate idle power [kVA] to effective power in [kW] (about 0.8)
CR	Corporate Responsibility
DC	Direct Current
DOD	Depth of Discharge
DG	Diesel Generator (Genset)
ICT	Information and Communication Technologies
KPI	Key Performance Indicator
MPP	Maximum Power Point
NPC	Net present Cost
OPEX	Operating Expenditure
OPzV(S)	Battery type (Ortsfest, Panzerplatten, Standard (S) or V Verschlossen
PSU	-48V Power Supply Unit
RBS	Radio Base Station (Site), 2G and 3G technology in the Radio Access Network
RES	Renewable Energy Sources
PR	Performance Ratio
SOC	State of Discharge
STC	Standard Test Conditions
TCO	Total Cost of Ownership
VRLA	Valve-regulated lead-acid
WS	Workshop
WTG	Wind turbine generator

6.3   Abbreviations (formulas)

Variable	Meaning	Unit
A	Autonomy time w/o DG	days
Cbat	Nominal battery capacity	Ah
Cbatah	Cost of battery capacity	€/Ah
Cchbd	CAPEX for bi-directional Charger	€ device
Cchbdkw	CAPEX for bi-directional Charger	€/kW
Cchsolkw	CAPEX for battery solar charger	€/kW
Cdgkva	Cost per Diesel generator	€/kVA of DG
Cpvkwp	Cost per kWp	€/kWp
Cpvwp	Cost of solar per Wp	€/Wp
Cwtgkw	Cost per turbine	€/KW of turbine
DoD	Depth of discharge	%
Epv	Annual energy from PV	kWh/kWp * a
Ewtg	Annual energy from Wind	kWh/a
i	Interest	%
inst	Installation + implementation fee	%
ks	Solar correction factor	-
LF	Load factor (efficiency) of battery	%
n	Considered lifetime à Assumed lifespan	years
Nwtg	Number of necessary turbines	amount
Obat%	OPEX per year in %	% from Invest
Obata	OPEX per year absolut	€/year
Ochbd%	OPEX for bidirectional charger	% of investment
Ochsol%	OPEX for battery solar charger	% of investment
Odg%	OPEX per year in %	% of investment
Odga	OPEX per year in absolute terms	€/year
Opv%	OPEX per year in %	% of investment
Opva	OPEX per year in absolute terms	€/year
Owtg%	OPEX per year in %	% of investment
Owtga	OPEX per year in absolute terms	€/year
Pdc	DC Load in BTS	W
Pdgn	Diesel nominal power type	KVA
Ppv	Required PV power	kWp
Pwtgn	Turbine nominal power type	KW
Q	Required Battery Capacity	Ah
SQM	Required sqm PV per kWp	sqm/kWp
Sstc	Solar Standard Test conditions	kW/m2
U	System Voltage	V
Vbat	Battery power	V

6.4 Figures

6.5 Tables

7   Appendix

7.1   Example of a 'Site questionnaire to plan RES based radio stations'

Renewable Energy Sources at
Radio Base Stations

Cell Site Questionare			

0   Site location coordinates
     Site height above sea level

S.l.	Description	Parameters	Equipments used
1	BTS Configuration with Power consumption (Watts) :		
2	Microwave power consumption (Watts) :		
3	Footprint of equipment to be installed :		
4	Other Equipment planned for the site (Watts) :		
5	Ambient Conditions throughout the year & regions - Max temperature & Average - Min Temperature & Average - Humidity		
6	Mains Power Availability per day :		
7	Mains Power Condition : - Fluctuations - Neutral Problem - Lightning & surges		
8	Mains Power Cost/ Unit :		
9	Average DG running/day : - DG Capacity used - single phase or three phase - Diesel consumption/ hour		
10	DG Maintenance cost Diesel Cost/ litre :		
11	Air Conditioners - Single Phase - Three Phase :		
12	Free cooling units :		
13	Batteries :		
14	PSU		

Ground plan to be attached   _______________________

Contact person   _______________________

Note: Grid quality questions are not relevant for off grid sites

7.2   Example of a 3-phase design

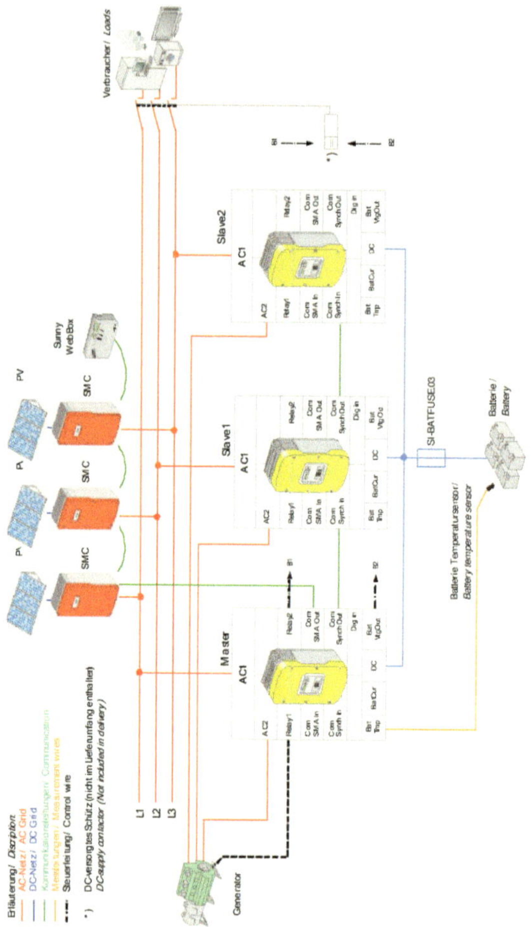

Figure 16: Example of a 3-phase layout with Sunny Island inverter (Source: SMA)

7.3   Example of an RES@RBS assessment with RAFORS

This is an example of an off-grid site in Albania which was assessed with RAFORS

To sum up:
| DC load: | AC Load:
| Typical 1.3kW at 48V | **1.8kW**
| + 0.1kW microwave system | Summer: 6h/d May – September
| + 0.1kW free cooling system | Winter: 3h/d October – April
| **= 1.5kW x 24h x 365d**

The mixed design was chosen due to the AC load (crosscheck: 12KVA sufficient to power AC and DC (3.3kW (about 4.2kVA) + 100% safety margin for battery charging and inrush current).

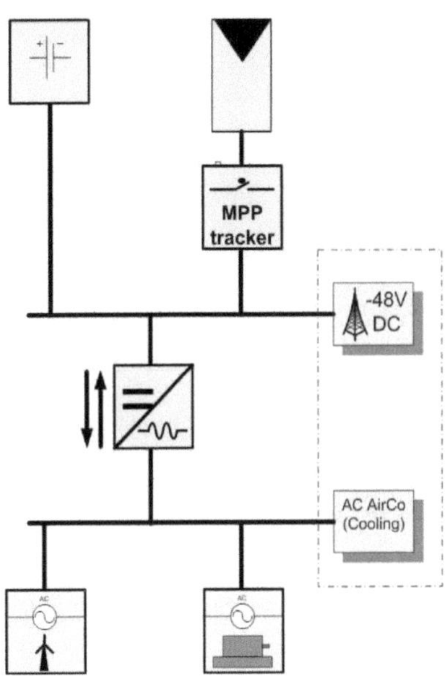

Due to the above mentioned conditions RAFORS uses the following data in its calculations:

DC Load in BTS	Pdc	1500 W		Interest	10% %
Location of the site (continent)	South Europe			Intsllation fee	10% %
Wind situation of the site (place)	Hill (75%)			Benchmark 'outsourcir	30000 €/year

The model shows that the 'solar and wind' solution is the most cost effective. Assuming an operational period of 6 years for all scenarios, the solution is more cost effective after 2.5 years.

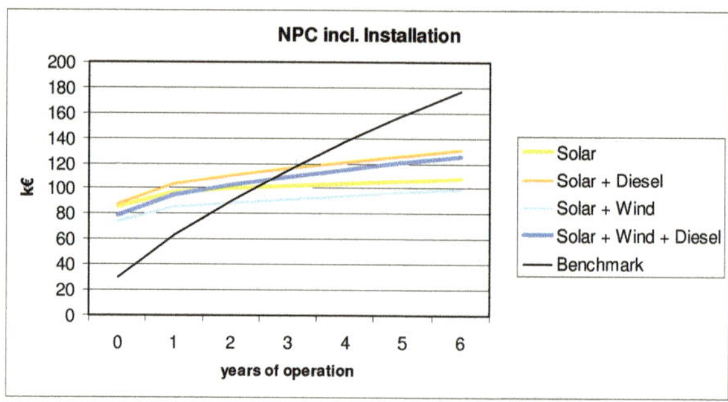

The following table summarises requirements based on the RAFORS output spreadsheet.

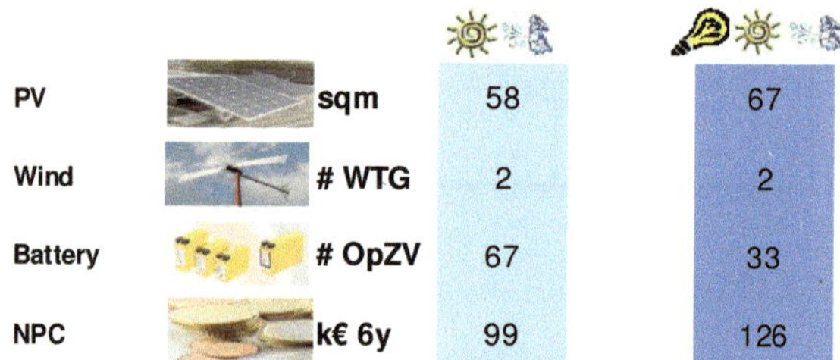

PV		**sqm**	58	67
Wind		**# WTG**	2	2
Battery		**# OpZV**	67	33
NPC		**k€ 6y**	99	126

Since the solar + wind scenario appears to be the most cost effective, the recommendation would be to start off initially with the GenSet configuration (dark blue) because of:
-   high temperatures at the location may give rise to active cooling requirements
-   the photo shows that the site is difficult to access, so longer autonomy time could be necessary
-   the significance of the site (two microwave links) as a transmission hub

Migration costs should therefore be considered in respect of:

-   increasing the number of batteries from 44 to 89
-   re-deployment (hopefully unnecessary) of the GenSet after 1 year

The example also shows how the running time of a diesel GenSet can be reduced by the use of batteries (section 6.3.9, [9])

7.4   Suppliers (examples)

This table provides readers with an initial overview of products and potential suppliers. It is not a product or supplier recommendation.

Product	Type / Name	Link to supplier	Picture
Solar Module	HIP-190 B2	www.sanyo.co.jp/clean/solar/hit_e/index_e.html	
Solar Module	Sunmodule SW 160…185	www.solarworld.de	
Battery Charger	Power Tarom 5055,..4120..4140	http://www.steca.com	
Bidirectional Battery Charger	Sunny Island SI 5048	www.sma.de	
Wind	BWC EXCEL 10KW	www.bergey.com	

7.5   Symbols (electrical)

 Batteries

 Solar Charger,
here with MPP tracker

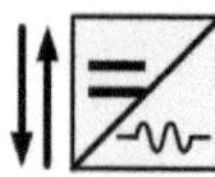 Stand alone Inverter

 PV generator (DC source)

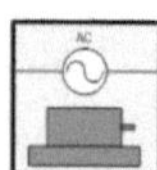 Diesel Generator (AC source)

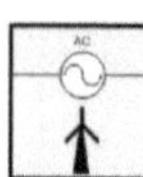 Wind turbine (AC source)

 Radio Base Station Load (DC)

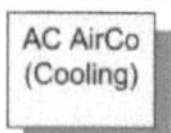 Radio Base Station Load (AC)

 Electrical connection